spot

BIG BIRDS

CRANES

by Lisa Amstutz

AMICUS | AMICUS INK

legs

bill

Look for these words and pictures as you read.

wings

nest

Look at that big bird.

It is a crane.

Some cranes are taller than you!

A crane has long legs.
It wades in the water.
It sleeps standing up.

legs

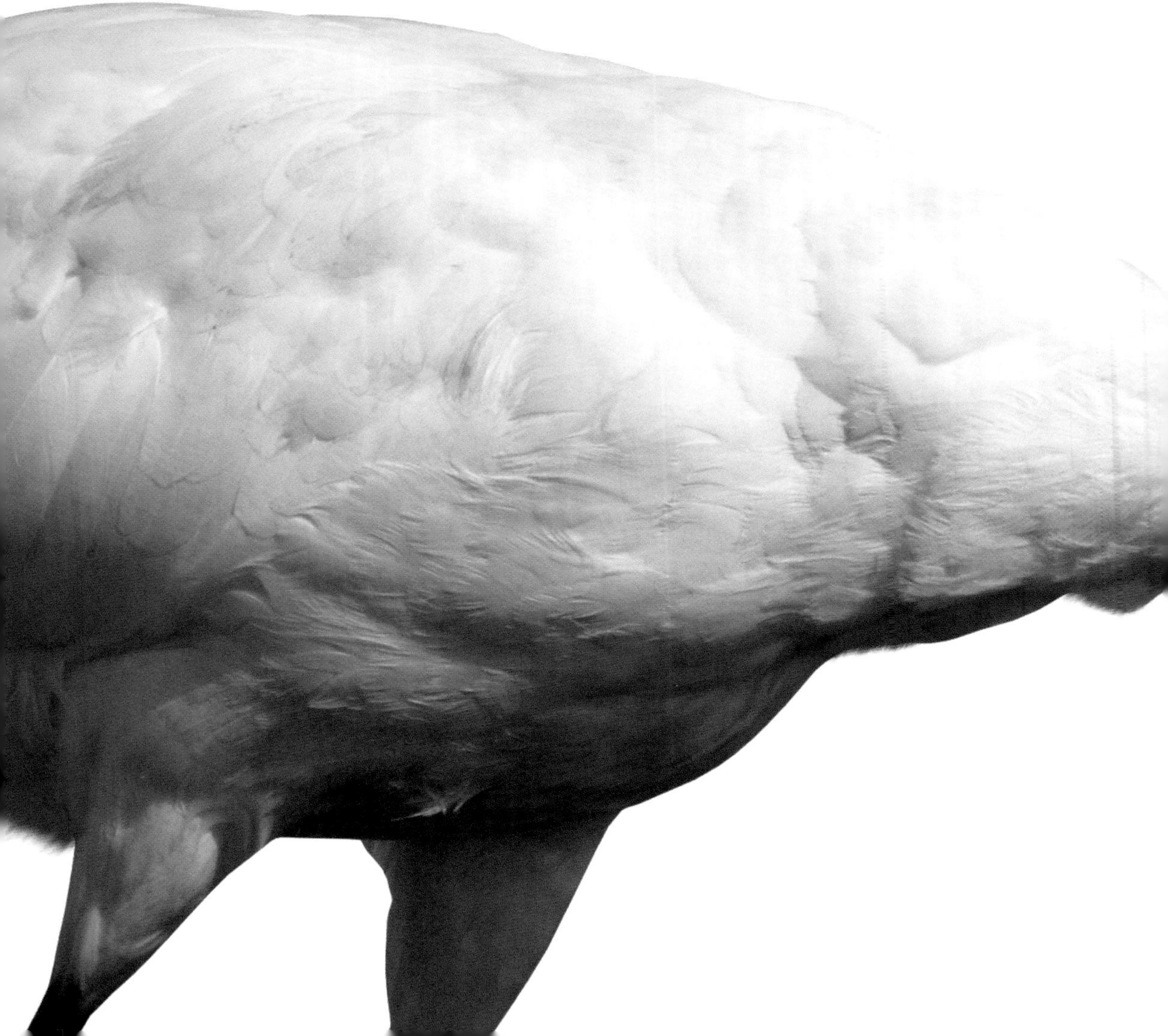

bill

A crane's bill is long, too.
Cranes eat fish.
They also eat bugs and plants.

wings

A crane flaps its big wings.
Cranes fly high.
Their long necks stick out.

Cranes dance to find mates.
They call.
They hop up and down.

Mom and Dad make a nest.
Mom lays two eggs.
The eggs have spots.

nest

The chicks hatch.

In a few months, they will fly!

legs

bill

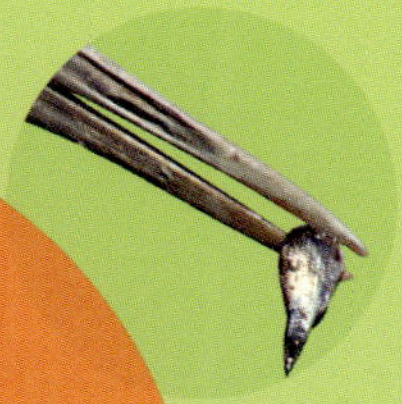

Did you find?

wings

nest

Spot is published by Amicus and Amicus Ink
P.O. Box 227, Mankato, MN 56002
www.amicuspublishing.us

Library of Congress Cataloging-in-Publication Data
Names: Amstutz, Lisa J., author.
Title: Cranes / by Lisa Amstutz.
Description: Mankato, MN : Amicus, [2023] | Series: Big birds | Audience: Ages 4-7 | Audience: Grades K-1 | Summary: "Majestic, long-legged cranes are showcased in this leveled reader about the big bird's habitat, body parts, and behavior. Includes a search-and-find feature to reinforce key vocabulary"-- Provided by publisher.
Identifiers: LCCN 2019060275 (print) | LCCN 2019060276 (ebook) | ISBN 9781645490968 (library binding) | ISBN 9781681526638 (paperback) | ISBN 9781645491385 (pdf)
Subjects: LCSH: Cranes (Birds)--Juvenile literature.
Classification: LCC QL696.G84 A47 2023 (print) | LCC QL696.G84 (ebook) | DDC 598.3/2--dc23
LC record available at https://lccn.loc.gov/2019060275
LC ebook record available at https://lccn.loc.gov/2019060276

Gillia Olson, editor
Deb Miner, series designer
Ciara Beitlich, book designer
Bridget Prehn, photo researcher

Photos by Shutterstock/Eric Isselee cover, 16; Shutterstock/athoms 1; Shutterstock/sichkarenko.com 3; iStock/gnagel 4–5; Getty/setimino 6–7; Shutterstock/Independent birds 8–9; Alamy/georgesanker.com 10–11; iStock/Melanie Coulthurst 12–13; iStock/jocrebbin 14–15